Name:

Find the Synonym

Directions: Circle the correct synonym for each underlined word.

1. The women were **surprised** when they heard the whole story.

A. scared   B. shocked   C. angry   D. ashamed

2. The class was **interesting**, but it was long.

A. boring   B. quiet   C. engaging   D. extensive

3. I have a dollhouse full of **miniature** furniture.

A. decorative   B. pretty   C. antique   D. tiny

4. That dirt road is winding and **bumpy**.

A. incline   B. rough   C. curvy   D. long

5. The social studies test was very **difficult**.

A. caring   B. challenging   C. easy   D. wise

6. I traded in my **old** bicycle for a new, shiny one.

A. used   B. ancient   C. sporty   D. crazy

7. We watched, amazed, as the deer **leapt** over the fence.

A. juggled   B. sauntered   C. ran   D. jumped

1

Name:

<u>Synonyms</u>

DIRECTIONS: Words that have the same or similar meanings are called synonyms. Circle

the word in each row that is the best synonym for the bold word.

1. **handsome**	pretty	attractive	tall
2. **explode**	combust	implode	spray
3. **turn**	spin	flat	swerve
4. **crazy**	insane	erratic	furious
5. **silly**	sane	goofy	stupid
6. **mistake**	accomplishment	effort	error
7. **scent**	smell	penny	aura
8. **talented**	educated	gifted	studied
9. **close**	near	squished	packed
10. **teach**	study	lecture	tutor

Recognizing Synonyms

Circle the correct synonym for each underlined word.

1. The children were not paying attention.

A. women   B. babies   C. kids   D. girls

2. Those wild girls are always getting into trouble.

A. rambunctious   B. quiet   C. loud   D. polite

3. The living room is very well lit this evening.

A. shining   B. white   C. dim   D. bright

4. Everyone in the stands cheered when Joe scored a goal.

A. whined   B. shouted   C. murmured   D. laughed

5. I asked the class to imagine that they were unicorns.

A. say   B. dream   C. pretend   D. wish

Read each row of words below. Circle the two words in each row that are synonyms.

6. fast	quick	old	zoom
7. eat	drink	watch	nibble
8. drive	ride	steer	draw
9. listen	jot	write	erase
10. iron	ear	dot	spot

Name: _____

 # **<u>Writing with Synonyms</u>**

DIRECTIONS: Using a dictionary or thesaurus, list at least two synonyms for each of the following common words. Then, on the line below, use one of each of the synonyms you chose in a sentence.

1. **bold** _____   _____

_____

2. **lie (noun)** _____   _____

_____

3. **flat** _____   _____

_____

4. **charming** _____   _____

_____

5. **blend** _____   _____

_____

6. **spry** _____   _____

_____

Name:

Synonyms

Circle the two words in each sentence that are synonyms.

1. The family with the noisy children is very loud.

2. Suddenly, the gang of boys stepped out of the crowd.

3. If this were a normal day, I would follow my regular schedule.

4. I dislike math because my mother hated it.

5. I will eat my dinner, then I will devour my dessert.

6. He sipped the soda, decided he liked it, and started to gulp.

7. I was in a bleak mood on that gloomy day.

8. Can you tell me the way to the path in the park?

9. My father saunters when he takes his walk.

10. My sister is kind to the animals because she is caring.

Read each row of words below. Circle the two words in each row that are synonyms.

11. sage	simple	dull	wise
12. mutter	talk	listen	hard
13. construct	appear	stature	assemble
14. packed	loose	skinny	compressed
15. morose	stoic	friendly	amiable

5

Name: _____

Serving Up Synonyms

Words that have the same or similar meanings are called synonyms. You can use a synonym to replace "tired" or overused words to make your writing more interesting.

DIRECTIONS: Replace each underlined word with a synonym. Write your word on the line.

1. When we arrived at the restaurant I was <u>starving</u>. _____

2. We were <u>shown</u> to a table by a window. _____

3. I <u>read</u> the menu very carefully; everything sounded so wonderful. _____

4. I finally <u>chose</u> to order the hamburger and fries. _____

5. Finally, the waiter <u>put</u> my dinner down in front of me. _____

6. I was so hungry that I <u>ate</u> everything very quickly. _____

7. By the time I was finished, I was <u>full</u>. _____

8. But then the waiter came to see if he could <u>lure</u> us with dessert. _____

9. I <u>like</u> sweets, so of course I had to select something. _____

10. It was the <u>best</u> dinner I have had in a long time! _____

Name: _____

Synonyms

Circle the two words in each sentence that are synonyms.

1. The boy kept trying to speak while I was talking.

2. The test was hard because the material is so difficult.

3. We played catch; I threw the ball and my dad tossed it back.

4. He hair is shiny and her teeth are bright.

5. One dog jumped up and the other leapt over the fence.

6. Listen closely so you will hear what I am saying.

7. I will find my wallet if I can locate my purse.

8. My brother is sad, and it makes me unhappy.

9. I lost my shoes and misplaced my hat.

10. I have two brothers and she has two siblings.

Write a synonym for each word.

11. snow _____

12. dirt _____

13. sky _____

14. shelf _____

15. porthole _____

16. ship _____

17. right _____

18. house _____

19. wage _____

20. school _____

7

Name: _____

Find the Synonyms

DIRECTIONS: Match the words on the left with their best synonyms on the right.

_____1. leaves	A)	dusk
_____2. evening	B)	make
_____3. morning	C)	brood
_____4. cook	D)	toil
_____5. prepare	E)	frolic
_____6. cut	F)	two
_____7. singing	G)	foliage
_____8. play	H)	drudgery
_____9. think	I)	bake
_____10. group	J)	caroling
_____11. pair	K)	dawn
_____12. work	L)	cover
_____13. boring	M)	gang
_____14. wrapper	N)	sheaf
_____15. papers	O)	slice

Perk Up Your Writing with Synonyms

Words that have the same or similar meanings are called synonyms. You can use a synonym to replace "tired" or overused words to make your writing more interesting.

DIRECTIONS: Read the paragraph below. Replace each bold word with a synonym that will add more detail and interest to the paragraph. Write your word on the correspondingly numbered line.

Mona was having a **bad**[1] day. First, she spilled her cereal on her **nice**[2] outfit. Then she had to **go**[3] to the bus stop in the rain. When she got to her first class, she was **unhappy**[4] to hear there was going to be a pop quiz. It was not a subject she was **good**[5] at, and she didn't do well. At lunch time, the cafeteria was only serving foods she did not **like**[6], and her best friend made her **unhappy**[7] because she didn't save her a seat at the lunch table. In the afternoon, Mona had gym class, which she **disliked**[8].

She thought her gym uniform was **ugly**[9] and none of her friends were in her class. After gym she felt **unhappy**[10], and she **spoke**[11] back to her history teacher and got sent to the principal's office. The principal's secretary called Mona's mother, and when Mona's mother got there, Mona could tell that she was not **happy**[12]. When she got home Mona's mother **told**[13] her that she was grounded. Mona went up to her room and **got**[14] on her bed. "Oh well, she thought to herself. "Tomorrow is another day!"

1. _____

2. _____

3. _____

4. _____

5. _____

6. _____

7. _____

8. _____

9. _____

10. _____

11. _____

12. _____

13. _____

14. _____

Name:

Synonyms

A synonym is a word that has the same or similar meaning as another word.

DIRECTIONS: Read each sentence below. Write a synonym on the line for each underlined word.

1. Our teacher was **happy** with our test scores. _____

2. The **loud** music got on Harriet's nerves. _____

3. Karen often **speaks** at conferences. _____

4. Rachel was **unhappy** to hear about our arrangement. _____

5. Only the **good** athletes can enter the race. _____

6. Mrs. Potter is a **good** mother. _____

7. We hope our new teacher is going to be **nice**. _____

8. The class was **long**. _____

9. What a **colorful** sweater. _____

10. That **thin** man is my neighbor. _____

11. It was a **good** meal. _____

12. Paul is good at building things. _____

13. Her decorations were very **fancy**. _____

14. Their **behavior** was getting on our nerves. _____

Name: _____

<u>Synonyms</u>

DIRECTIONS: Read each set of words below. Identify whether they are synonyms or antonyms. Write synonyms or antonyms on the line.

_____1. bravery, courage

_____2. bold, timid

_____3. succeed, triumph

_____4. wet, damp

_____5. bright, dim

_____6. ask, inquire

_____7. dislike, enjoy

_____8. store, shop

_____9. silent, quiet

_____10. easy, difficult

DIRECTIONS: Find a synonym for each word below. Write it on the line.

<u>SYNONYM</u>

11. tasty _____

12. loud _____

13. chubby _____

14. think _____

15. confuse _____

16. eat _____

17. argue _____

<u>SYNONYM</u>

18. cool _____

19. hard _____

20. winner _____

21. paint _____

22. gigantic _____

23. fly _____

24. sleep _____

Synonyms

Words that have the same or similar meanings are called synonyms. You can use a synonym to replace "tired" or overused words to make your writing more interesting.

DIRECTIONS: Replace each underlined word with a synonym.

Choose from the words in parentheses. Write your word on the line.

1. Mr. Martin enjoys hiking up Mount Williams. _____

   (running, trekking, sauntering)

2. My father always sleeps late on Sundays. _____

   (snoozes, lays, sits)

3. Our class is responsible for refreshments. _____

   (accountable, believable, reasonable)

4. My Aunt Louise was a remarkable person. _____

   (amazing, boring, unimpressive)

5. I thought it was a magnificent play. _____

   (awful, gigantic, fantastic)

6. Ellie is very keen on gardening. _____

   (ungrateful, dislike, eager)

7. I find it hard to respect Ms. Montgomery. _____

   (believe, admire, dislike)

Name:

Find the Synonyms

Words that have the same or similar meanings are called synonyms.

DIRECTIONS: Read the word at the beginning of each line.

Color the boxes of the words that are its synonyms.

1. talent	flair	aptitude	failure
2. limber	stiff	agile	athletic
3. enthusiastic	avid	dull	excited
4. regret	appreciate	rue	repent
5. prevent	avert	stop	deter
6. inspiration	muse	stuck	boring
7. compete	consistent	vie	cooperate
8. struggle	wrangle	fight	agree
9. sleek	glossy	lustrous	rough
10. peaceful	calm	serene	shocking
11. wise	sage	learned	aged

Name:

<u>Synonym Knowledge Check</u>

1. What is a synonym?

_____

_____

2. What is one reason to use synonyms in your writing?

_____

_____

3. Circle the word that is the best synonym for **accident.**

trip   calamity   purposeful   fall

Write a synonym for each of the words below.

good   _____   mad   _____

nice   _____   bad   _____

kind   _____   light   _____

happy   _____   dark   _____

unhappy   _____   person   _____

walk   _____   talk   _____

Name:

<u>Synonyms</u>

DIRECTIONS: Write a synonym for each word.

1. walk _____

2. warm _____

3. scare _____

4. drink _____

5. draw _____

6. skinny _____

7. watch _____

8. smelly _____

9. grin _____

10. break _____

11. trip _____

12. tired _____

13. small _____

14. dirty _____

15. angry _____

16. filthy _____

17. smart _____

18. caring _____

19. pleasant _____

20. awesome _____

21. huge _____

22. need _____

23. respect _____

24. cheap _____

25. current _____

26. known _____

27. great _____

28. sharp _____

Name:

Synonyms

Directions: Write each word next to its synonym.

leap	handsome	bad	icy
messy	hard	fast	easy
right	small	noisy	present
stone	friend	angry	toss

quick		loud	
mad		rock	
buddy		throw	
dirty		awful	
simple		tiny	
correct		cute	
difficult		cold	
gift		jump	

Name:

Synonyms

Synonyms are words that have the same meaning. Synonyms for **big**: large, huge, gigantic.

Circle the 2 synonyms for each set of words.

1.	tasty	chilly	cool	comfortable
2.	walk	crawl	stroll	run
3.	scare	confuse	frighten	argue
4.	bravery	honesty	courage	winner
5.	record	paint	draw	sketch
6.	cook	drink	eat	munch

Write a synonym for each word.

7. **thin** _____   8. **tiny** _____

9. **store** _____   10. **fall** _____

11. **tasty** _____   12. **see** _____

13. **silent** _____ 14. **friendly** _____

15. **yell** _____   16. **smile** _____

17. **think** _____   18. **run** _____

19. **jump** _____   20. **nap** _____

Name: _____

Synonyms

Read each sentence. Decide which word in the box means almost the same as the underlined word. Then write it on the line.

small	nap	pal	large	leap
glad	road	shout	sick	stone

_____   1.   I ate the **big** apple.

_____   2.   The baby took a **rest**.

_____   3.   Please don't **yell**.

_____   4.   I saw a **little** bunny.

_____   5.   I threw a **rock** in the lake.

_____   6.   I am so **happy** you came.

_____   7.   I really feel **ill** today.

_____   8.   You are my best **friend**.

_____   9.   A frog can **jump** very far.

_____   10.   I ran down the **street**.

Name: _____

Synonyms

children	tale	damp
done	giant	like
pebble	silly	chuckle

Choose a synonym from the box to replace each underlined word.

1. Tommy liked watching the **huge** elephant at the zoo. _____

2. I tossed a **stone** in the lake. _____

3. Carla knows so many **funny** jokes. _____

4. Will you tell me a **story**, Grandpa? _____

5. Sarah's bedroom is very **neat**. _____

6. Ed makes everyone **laugh** when he makes goofy faces. _____

7. I **enjoy** drinking iced tea during the summer. _____

8. The **kids** at the park played basketball. _____

9. Put the **wet** towel on the clothesline to dry. _____

10. Put your plate in the sink when you're **finished**. _____

11. **Maybe** we can go outside after lunch today. _____

12. You should **start** your science project tonight. _____

Synonyms

Directions: Write each word in the box next to its synonym.

Finish	End	
Journey	Trip	
Lamp	Memo	
Rich	Mist	
Money	Noon	
Fog	Wealthy	
Midday	Cash	
Note	Light	

Name:

Synonyms

Write a synonym for each of the underlined words.

1. The baby was **<u>sobbing</u>** loudly.   _____

2. I enjoyed the **<u>thrilling</u>** movie.   _____

3. My classmate is very **pretty**.   _____

4. There is a **<u>tiny</u>** spider on the wall.   _____

5. Susan thought her homework was too **<u>hard</u>**.   _____

6. I should eat a **<u>big</u>** lunch because I am starving.   _____

Decide if the following pair of words are synonyms. Write yes or no.

1. fat / thin   _____

2. quickly / speedily   _____

3. huge / massive   _____

4. junior / senior   _____

5. noisily / loudly   _____

Name:

<u>Synonyms</u>

Write a synonym for the underlined words.

1. The **<u>big</u>** bear slept quietly. _____

2. Her eyes were **<u>pretty</u>**. _____

3. Tom's house was **<u>dirty</u>**. _____

4. The performance was **<u>great</u>**. _____

5. His shoes were **<u>small</u>**. _____

6. Joshua's dad was **<u>nice</u>**. _____

7. Nylah was **<u>grumpy</u>**. _____

Name: _____

Replacing Words with Synonyms

Synonyms are words that have the same meanings.

Synonyms for beautiful: pretty, gorgeous, attractive

below	glad	ill	hilarious	present	bright
error	thief	plump	blend	clean	aqua

Replace each underlined word with a synonym from the word box.

1. His uncle brought a birthday **gift** wrapped in red paper. _____

2. The **fat** cat could barely climb up the tree. _____

3. My teacher was **happy** to see me. _____

4. The painting had **brilliant** shades of blue. _____

5. The dog curled up **under** the table. _____

6. My dad knows a lot of **funny** jokes. _____

7. The **burglar** stole all the diamonds and gold. _____

8. I went home early because I felt **sick**. _____

9. You will need to **mix** the eggs and flour. _____

10. The **cost** of the toy was to high. _____

11. The fisherman stared out at the **blue waters**. _____

12. Maxie made a **mistake** on his math test. _____

Synonym Words

full - packed

furious - enraged

future - coming

gain - acquire

gallant - chivalrous

gaudy - showy

gaunt – scrawny

generous - giving

gentle - mild

genuine - authentic

gigantic - immense

glad - happy

gloomy - dark

good - nice

great - outstanding

handy - useful

hard - firm

hate - loathe

help - aid

high - lofty

hold - grip

honest - sincere

hospitable - welcoming

huge - vast

humble - modest

humiliate - embarrass

identical - alike

idle - inactive

ignorant - uninformed

immaculate - pure

immune - resistant

impartial - neutral

impatient - eager

imperative - mandatory

imperfect - marred

jolly - merry

Thank You for purchasing. Hope you have a wonderful day.

- Kandys Bookstore